T0413230

SEEDS WE EAT

Katherine
Rawson

Creating Young Nonfiction Readers

EZ Readers lets children delve into nonfiction at beginning reading levels. Young readers are introduced to new concepts, facts, ideas, and vocabulary.

Tips for Reading Nonfiction with Beginning Readers

Talk about Nonfiction
Begin by explaining that nonfiction books give us information that is true. The book will be organized around a specific topic or idea, and we may learn new facts through reading.

Look at the Parts
Most nonfiction books have helpful features. Our *EZ Readers* include a Contents page, an index, and color photographs. Share the purpose of these features with your reader.

Contents
Located at the front of a book, the Contents displays a list of the big ideas within the book and where to find them.

Index
An index is an alphabetical list of topics and the page numbers where they are found.

Photos/Charts
A lot of information can be found by "reading" the charts and photos found within nonfiction text. Help your reader learn more about the different ways information can be displayed.

With a little help and guidance about reading nonfiction, you can feel good about introducing a young reader to the world of *EZ Readers* nonfiction books.

Mitchell Lane
PUBLISHERS

2001 SW 31st Avenue
Hallandale, FL 33009
www.mitchelllane.com

First Edition, 2021.

Author: Katherine Rawson
Designer: Ed Morgan
Editor: Morgan Brody

Names/credits:
Title: Seeds We Eat / by Katherine Rawson
Description: Hallandale, FL :
Mitchell Lane Publishers, [2021]

Series: Plant Parts We Eat
Library bound ISBN: 978-1-883845-04-9
eBook ISBN: 978-1-58415-101-2

EZ readers is an imprint of Mitchell Lane Publisher

Photo credits: Freepik.com, Shutterstock

Contents

Plants make seeds.
The seeds grow into new plants.
Seeds have different shapes and sizes.

Did You Know?

There are hundreds of different kinds of beans in the world.

Many seeds are good to eat.
Beans and peas are seeds.
They grow inside **pods**.
The pods grow on a **vine**.

We eat some beans **fresh**.
String beans and lima beans
are fresh beans.

We eat some beans dried. Black beans, red beans, and mung beans are dried beans.

Mung Beans

Corn **kernels** are seeds.
The kernels grow on a cob.

Did You Know?

One corn cob has about 800 kernels.

Corn on the cob is fun to eat.
Popcorn is fun to eat, too.

We make flour from dried corn.
We use it to bake bread and muffins.

DID YOU KNOW?

Beans and nuts have a lot of protein. Protein is important for strong, healthy bodies.

Walnuts and almonds are seeds.
They grow on trees.
They make good snacks.
They taste good in cakes and cookies, too.

Did You Know?

California grows more almonds than any other place in the world. Over two million tons of almonds are grown there every year.

Did You Know?
We eat many kinds of seeds.
Flax
Sunflower
Sesame

We enjoy eating many kinds of seeds.

Glossary

fresh
Green, newly picked

kernel
Seed of a grain plant

pod
Part of a plant that holds seeds inside

vine
A climbing stem that cannot support itself

Sources

https://www.ncbi.nlm.nih.gov/pmc/articles/PMC3257681/

https://www.ag.ndsu.edu/publications/food-nutrition/all-about-beans-nutrition-health-benefits-preparation-and-use-in-menus

http://beaninstitute.com/bean-nutrition-overview/

https://www.hsph.harvard.edu/nutritionsource/what-should-you-eat/protein/

http://www.berkeleywellness.com/healthy-eating/food/article/types-fresh-beans

https://www.factmonster.com/encyclopedia/ecology/botany/general/nut

https://www.worldatlas.com/articles/the-top-walnut-producing-countries-in-the-world.html

https://www.iowacorn.org/education/faqs

https://www.worldatlas.com/articles/top-almond-producing-countries.html

Further Reading

Web Pages:
Learn about different kinds of seeds
https://www.massaudubon.org/content/download/8289/146629/file/seeds.pdf

Watch a video about how a bean grows
https://www.dkfindout.com/us/video/animals-and-nature/how-bean-grows/

Learn facts about corn
http://www.sciencekids.co.nz/sciencefacts/food/corn.html

Books:
How a Seed Grows
By Helene J. Jordan (Harper Collins, 2015)

A Seed Is Sleepy
By Dianna Aston (Chronicle Books, 2014)

A Bean's Life Cycle (Capstone Press, 2017)

Index

About the Author

Katherine Rawson loves growing, cooking, and eating vegetables. She also loves writing. So, she thought it would be a great idea to write books about plants we eat. She likes eating all kinds of beans. She also enjoys growing beans in her garden because the vines are so pretty.